ANIMALS THAT FIGHT

THINGS ANIMALS DO

Kyle Carter

The Rourke Book Co., Inc.
Vero Beach, Florida 32964

© 1995 The Rourke Book Co., Inc.

All rights reserved. No part of this book may be reproduced or utilized in any form or by any means, electronic or mechanical including photocopying, recording or by any information storage and retrieval system without permission in writing from the publisher.

Edited by Sandra A. Robinson and Pamela J.P. Schroeder

PHOTO CREDITS
All photos © Kyle Carter

Library of Congress Cataloging-in-Publication Data

Carter, Kyle, 1949-
    Animals that fight / by Kyle Carter.
      p. cm. — (Things animals do)
    Includes index.
    ISBN 1-55916-113-2
      1. Animal fighting—Juvenile literature. [1. Animal fighting.]
I. Title. II. Series: Carter, Kyle, 1949-   Things animals do.
QL758.5.C37 1995
591.51—dc20
                                            94-46854
                                                CIP
                                                 AC

Printed in the USA

TABLE OF CONTENTS

Fighting	5
Food Fights	6
Fishing Rights Fights	9
How Fights End	11
How Animals Fight	14
Dominance Fights	16
Bighorn Battles	19
Dueling Elk	20
Beach Masters	22
Glossary	23
Index	24

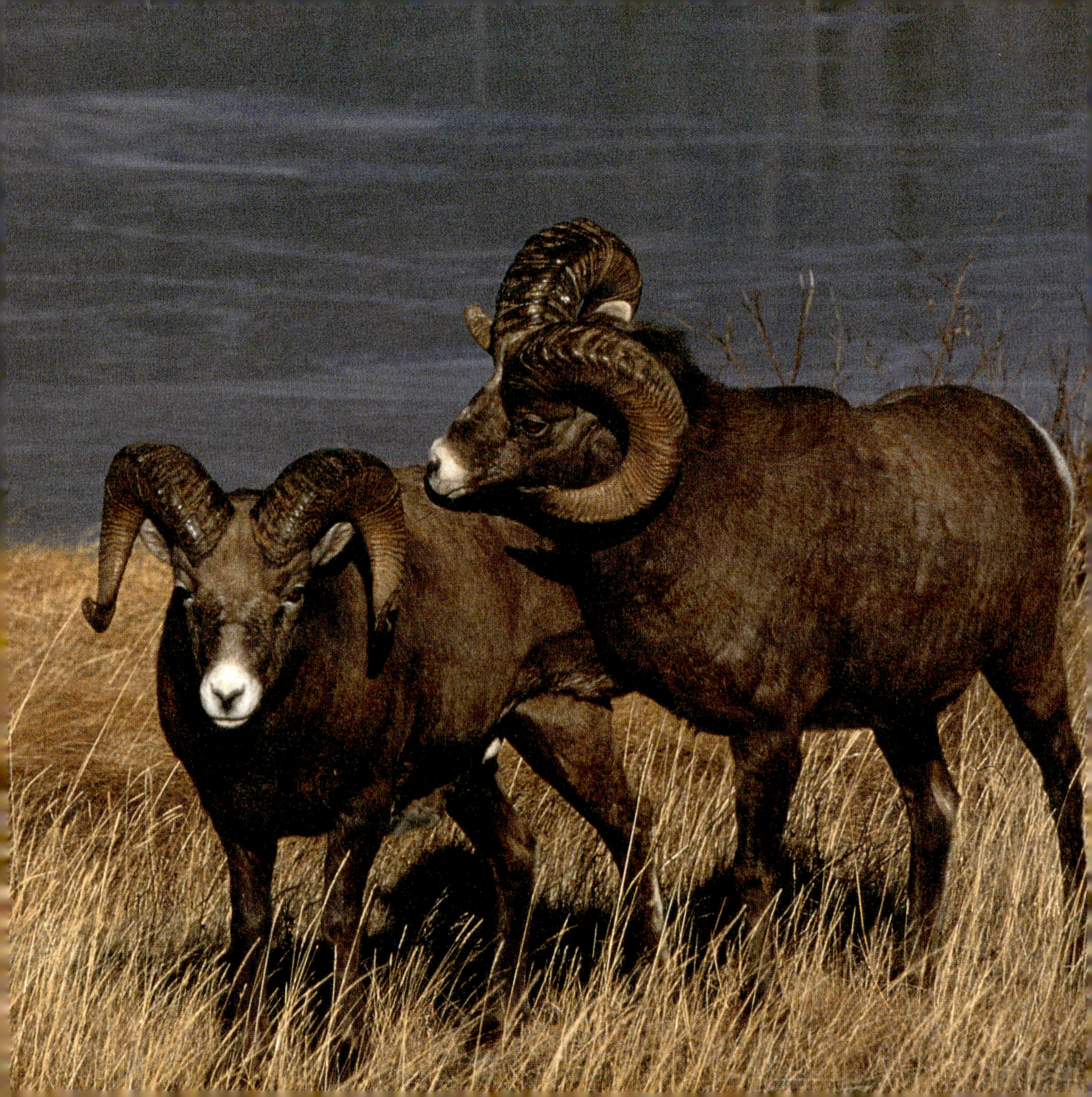

FIGHTING

Animals usually fight for one of two reasons—food or mates. Animals of different **species,** or kinds, may fight over food. Animals of the same species fight for the right to take mates.

Animals that fight risk injury or even death. They also burn energy that may be hard for them to replace.

Many animals have ways to avoid fighting. Often, an animal's scent, body language or call will scare an **opponent** away.

With a kick to another ram's belly, a bighorn signals that he wants to fight

FOOD FIGHTS

Animals do not always avoid fighting, of course. **Predators,** or meat-eating animals, have to attack and kill other animals for food. This type of "food fight" between a predator and its **prey** is lopsided. The attacker usually wins. Sometimes, though, a hungry predator bites onto more than it can chew, and its meal escapes.

Food fights also happen when one predator decides to pirate, or steal, food from another. Bald eagles often steal fish from each other and from ospreys.

With a strangling bite, a lioness kills a wildebeest while avoiding dangerous hooves and horns

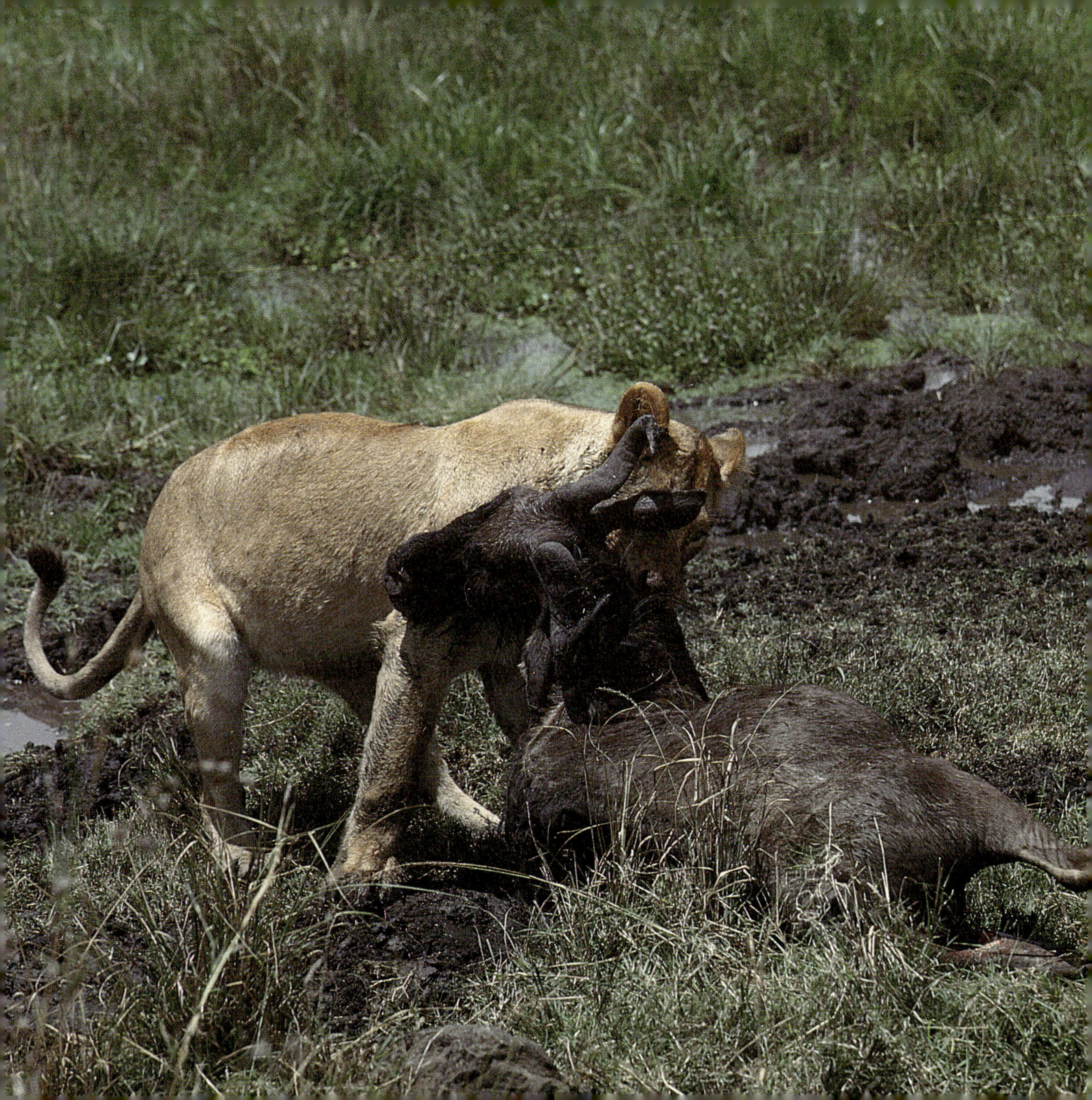

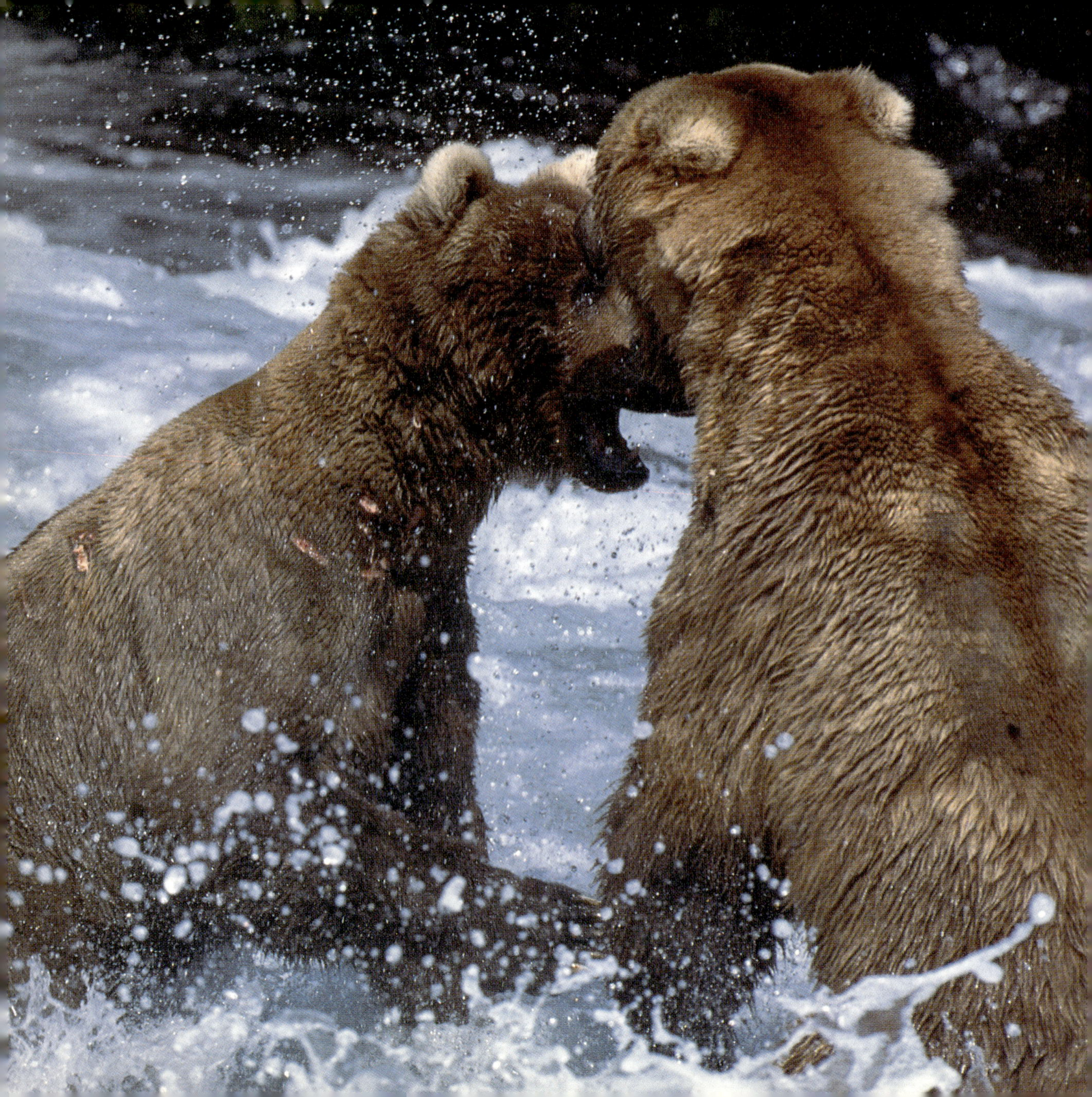

FISHING RIGHTS FIGHTS

Brown bears learn quickly where the best fishing is by watching other bears. The biggest bears always try to take the best fishing holes. Large brown bears often pirate salmon from smaller brown bears.

Big "brownies" easily bully smaller bears away. A bloody fight may start, though, if a large bear picks on another bear its own size.

These are really fights about food, not space. The winner of the space—the fishing hole—also wins the most food.

Two brown bears fight tooth and claw for fishing rights on an Alaskan stream

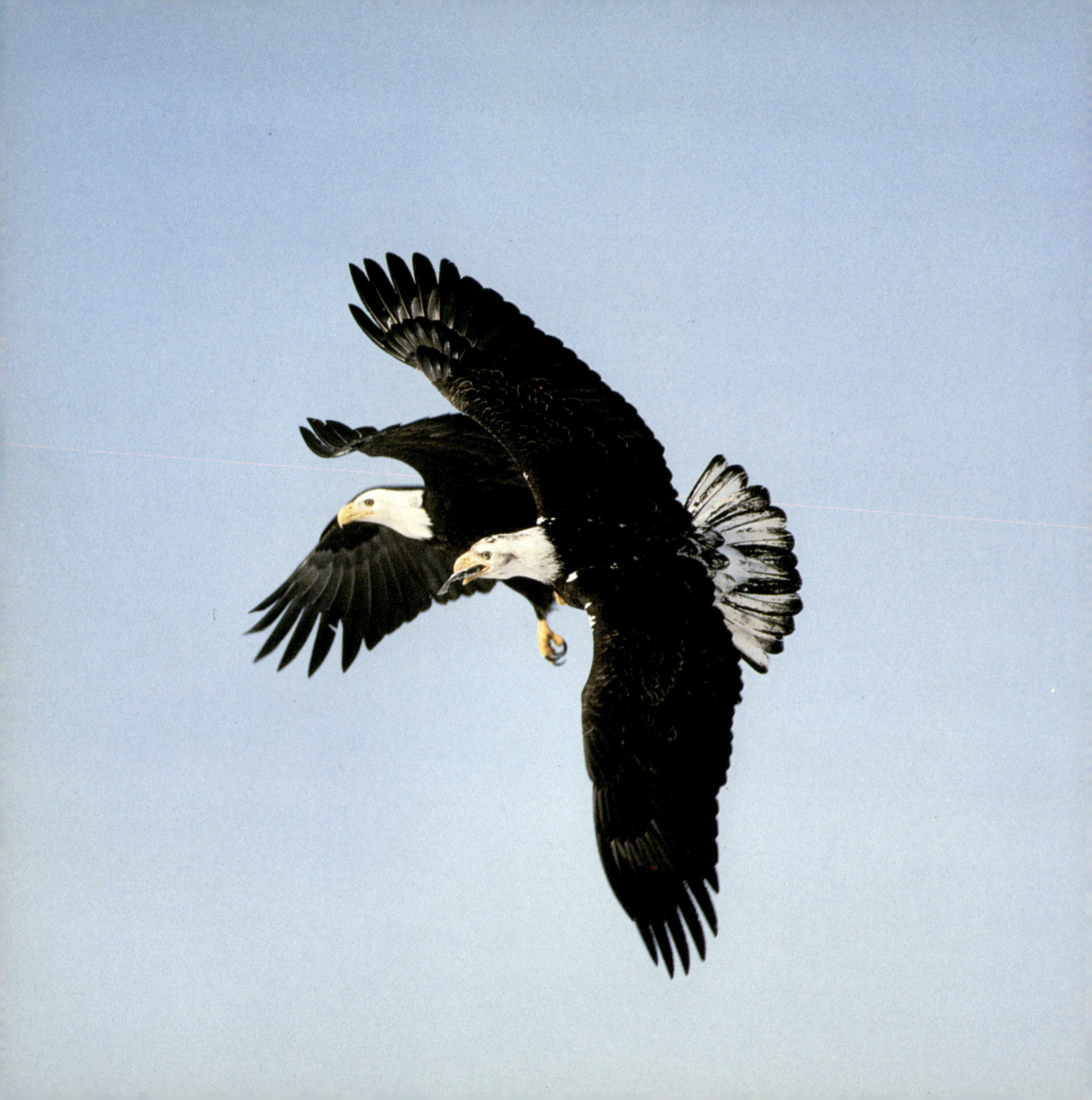

HOW FIGHTS END

Brown bears are big and strong enough to hurt each other badly. Like most animal fights, however, their battles usually end quickly and without serious injury.

Animals of the same kind rarely kill each other. That would be wasteful of the species. When one fighter quits, the other usually lets it walk or run away.

This midair fight between bald eagles for a fish was bloodless, but needle-sharp talons can cause serious injury

Male sockeye salmon fight over spawning beds in the shallows of British Columbia's Adams River

Grizzlies fight with fearsome fangs and claws

HOW ANIMALS FIGHT

All kinds of animals fight, from insects to large mammals. Each animal fights with the weapons it has.

Predators use their sharp beaks, teeth and claws. Large plant-eating animals fight each other with horns or antlers. Their headware also helps them fight back against predators.

Some birds, such as prairie chickens, have a hidden weapon—a sharp spur behind each foot. King penguins slap each other with their flippers.

Male prairie chickens are a fighting blur on a March dawn

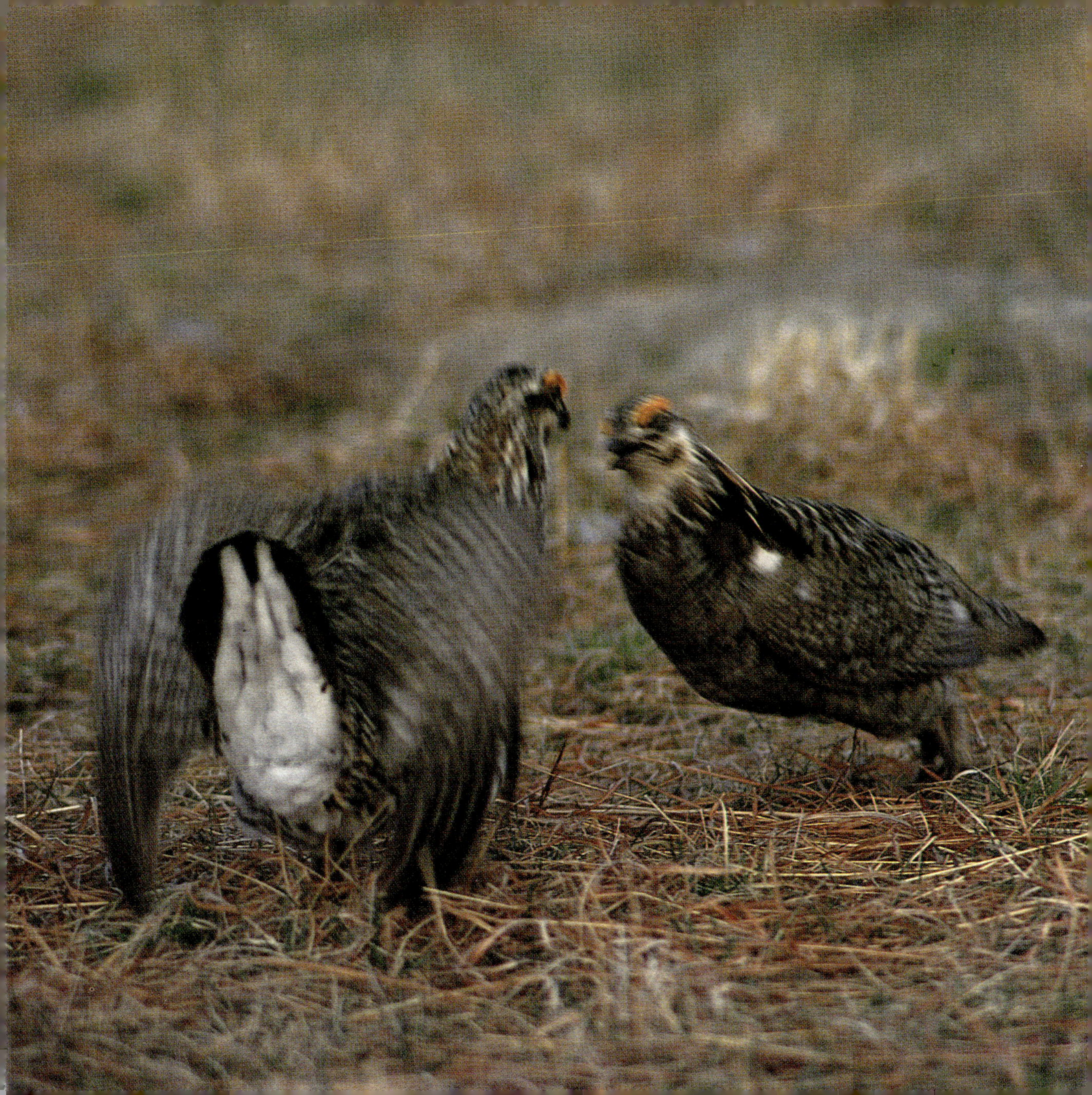

DOMINANCE FIGHTS

Instinct tells male animals of some species to fight each other during a certain season each year. These **dominance** fights decide which animal is the strongest, or most dominant.

The strongest male uses force to keep other males away from most of the females. The strongest male can then father most of the young animals. A strong father is good for the group. He helps produce strong, healthy young.

Fighting for control of females, bull elephant seals battle in the California surf

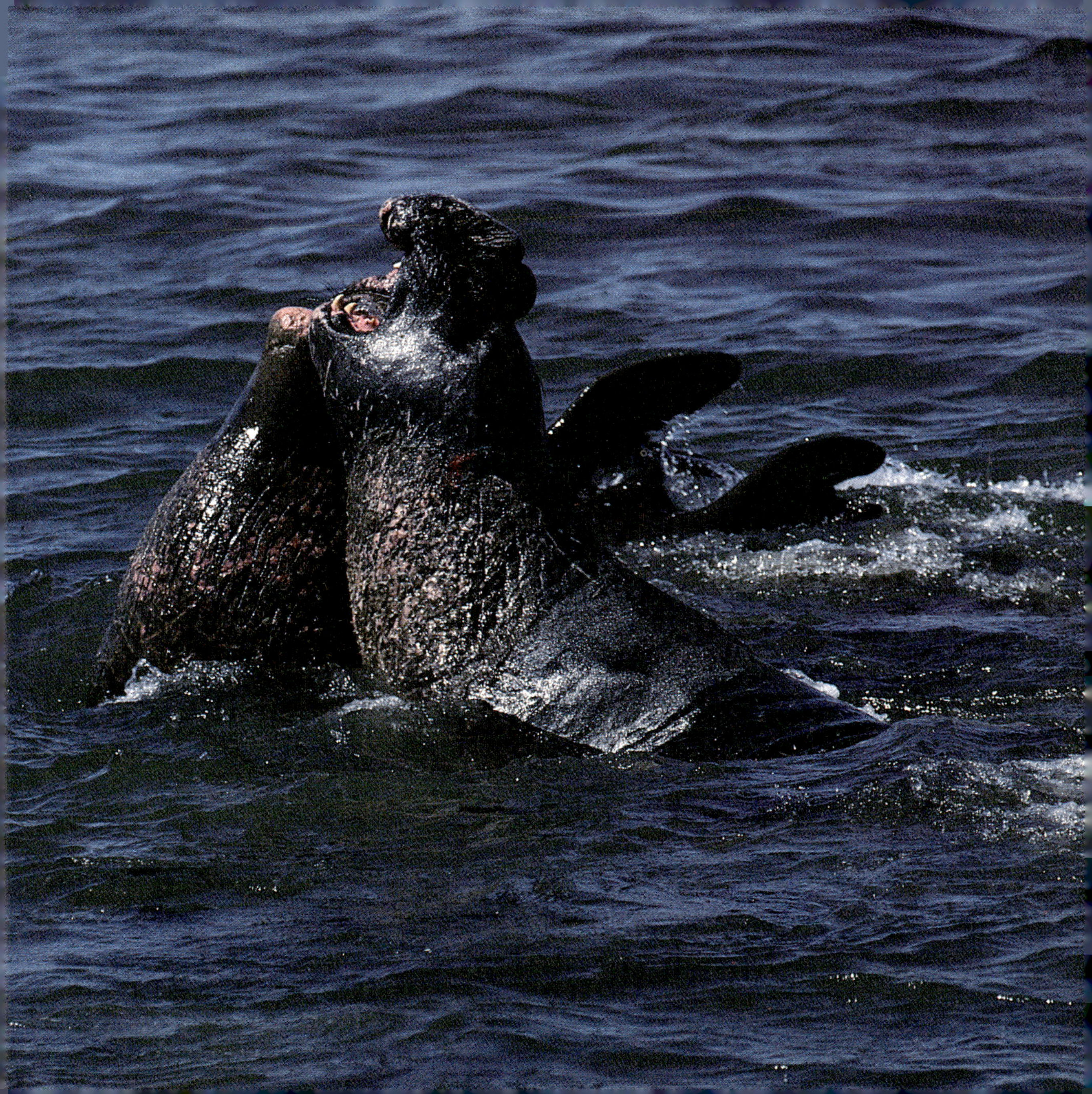

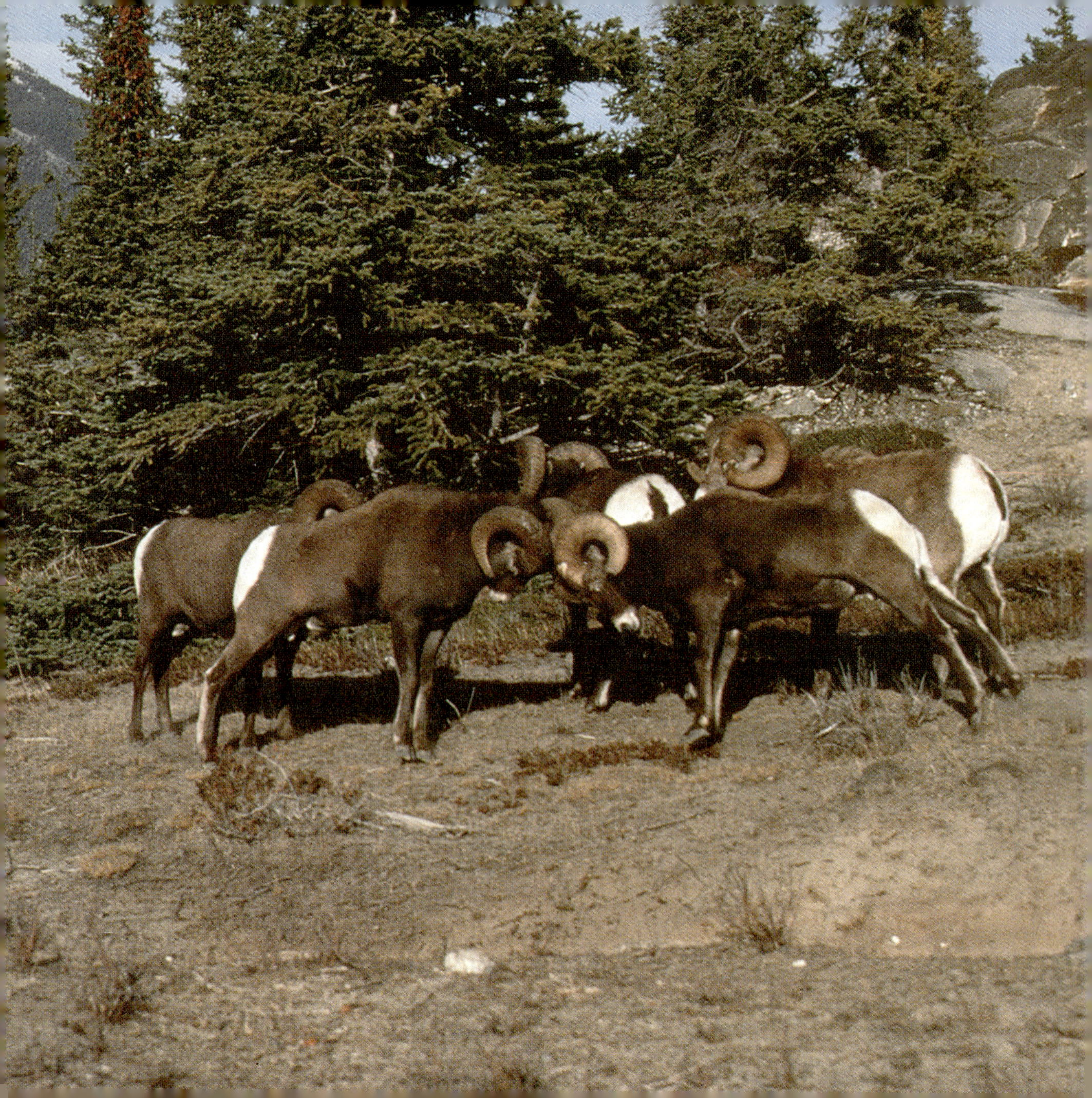

BIGHORN BATTLES

Bighorn sheep rams are built to fight each other. They have thick, curled horns and skulls with two layers of bone for protection.

Rams continually test their strength in butting contests. Most of the serious, head-jarring duels happen each fall when the rams are seeking mates.

The sheep's horns are for battling the horns of other rams. Bighorns are not a threat to other animals unless they're attacked.

Jarring each other with head-popping whacks, bighorn rams duel in the Rocky Mountains of Canada

DUELING ELK

During most of the year, bull elk quietly graze and rest. However, each fall, bull elk show off a new rack of antlers and a new mood. They are loud, ornery and ready to fight.

A bull elk's high-pitched bugle call warns other bulls to stay away. If another bull ignores the warning, he may find himself in a head-to-head duel with the meadow's master.

What's the prize for the dominant bull? He wins the herd of cow elk that stand nearby.

Bull elk battle for the right to control a herd of cow elk

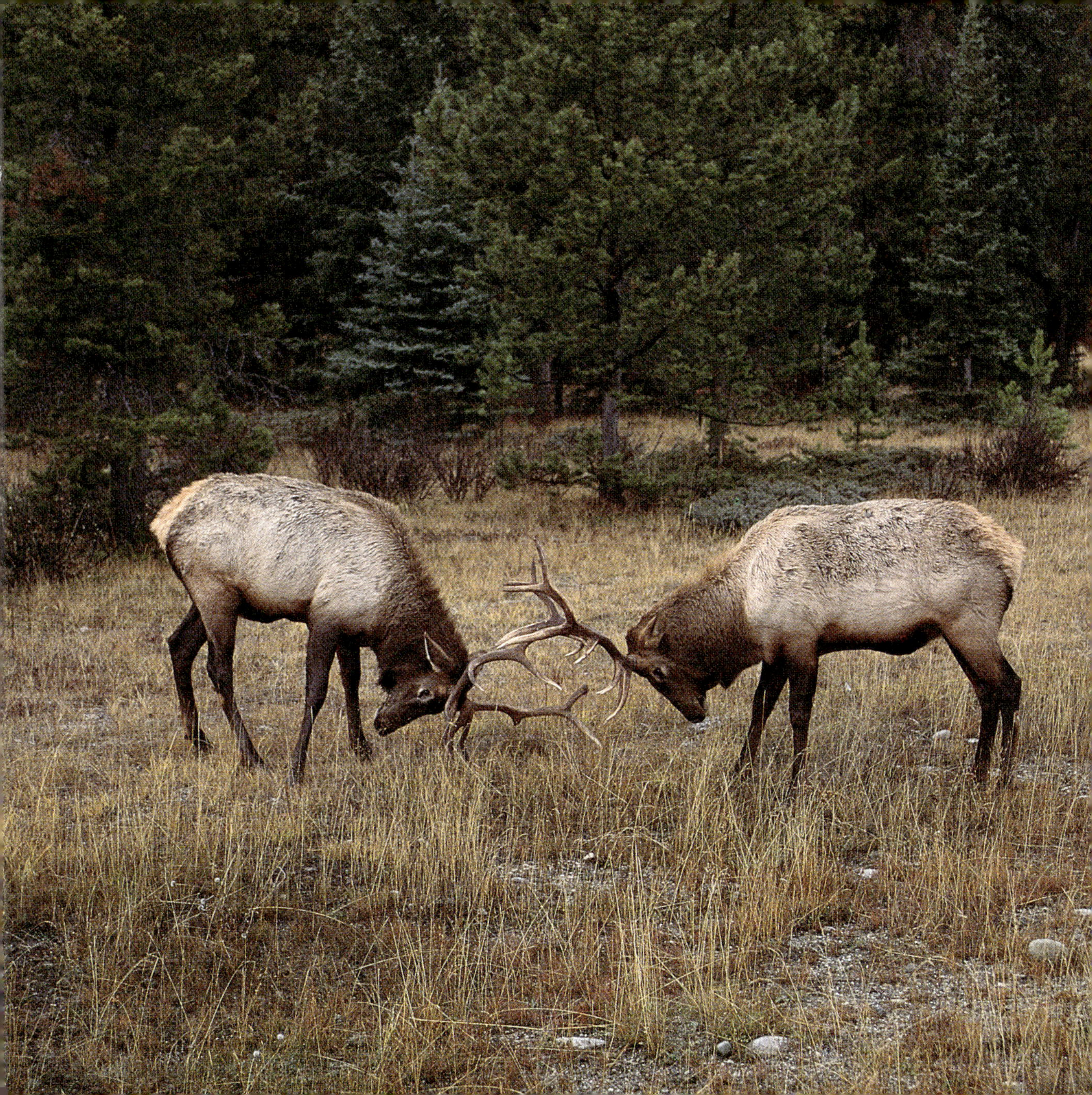

BEACH MASTERS

Elephant seal bulls are brownish-gray mountains of muscle and fat. A bull can be 18 feet long and weigh over 8,000 pounds—more than two cars!

Elephant seal bulls fight for mates. They belly-up to each other and bite with long, sharp teeth. Elephant seal fights can be bloody, but the animals' fat usually saves them from being badly hurt. A fight may last 30 minutes, or until one bull backs off.

Glossary

dominance (DAH min entz) — power over others

instinct (IN stinkt) — things an animal knows how to do without being taught

opponent (o PO nent) — the enemy; the one that fights another in battle

predator (PRED uh tor) — an animal that kills another animal for food

prey (PRAY) — an animal that is hunted by another animal for food

species (SPEE sheez) — within a group of closely-related animals, one certain kind, such as a *grizzly* bear

INDEX

antlers   14, 20
bears, brown   9, 11
eagles   6
elk   20
fighting   5, 6
fights   9, 11, 22
   dominance   16
   food   6
food   5, 6, 9
horns   14, 19
insects   14
language, body   5
mammals   14
mates   5, 19, 22
ospreys   6
prairie chickens   14
predators   6, 14
prey   6

salmon   9
scent   5
seals, northern elephant   22
sheep, bighorn   19